MY FIRST SCIENCE BIOGRAPHY

Marie Curie

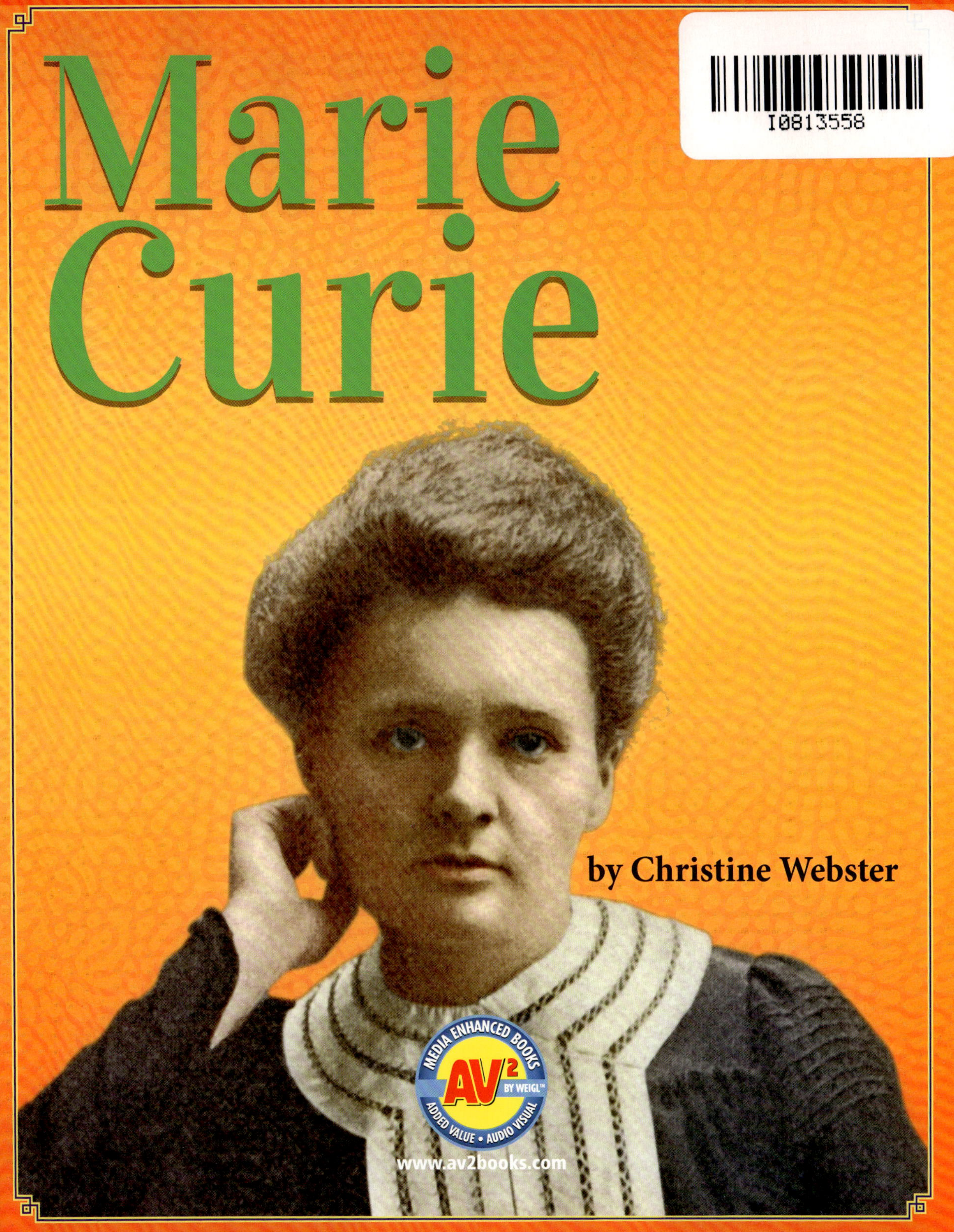

by Christine Webster

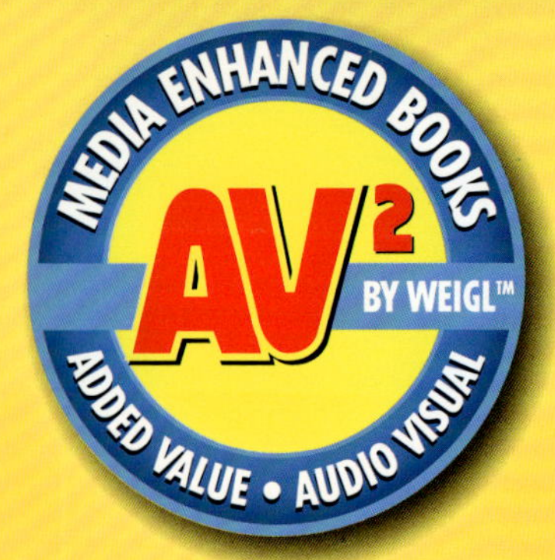

Go to **www.av2books.com**, and enter this book's unique code.

BOOK CODE

AVR35944

AV² by Weigl brings you media enhanced books that support active learning.

AV² provides enriched content that supplements and complements this book. Weigl's AV² books strive to create inspired learning and engage young minds in a total learning experience.

Your AV² Media Enhanced books come alive with...

Audio
Listen to sections of the book read aloud.

Key Words
Study vocabulary, and complete a matching word activity.

Video
Watch informative video clips.

Quizzes
Test your knowledge.

Embedded Weblinks
Gain additional information for research.

Slideshow
View images and captions, and prepare a presentation.

Try This!
Complete activities and hands-on experiments.

... and much, much more!

Published by AV² by Weigl
350 5th Avenue, 59th Floor
New York, NY 10118
Website: www.av2books.com

Library of Congress Control Number: 2019938597

ISBN 978-1-7911-0954-7 (hardcover)
ISBN 978-1-7911-0955-4 (softcover)
ISBN 978-1-7911-0956-1 (multi-user eBook)
ISBN 978-1-7911-0957-8 (single-user eBook)

Printed in Guangzhou, China
1 2 3 4 5 6 7 8 9 0 23 22 21 20 19

062019
311018

Project Coordinator: Heather Kissock
Art Director: Terry Paulhus

Photo Credits
Every reasonable effort has been made to trace ownership and to obtain permission to reprint copyright material. The publishers would be pleased to have any errors or omissions brought to their attention so that they may be corrected in subsequent printings.

Weigl acknowledges Getty, Alamy, Newscom, Shutterstock, and Wikimedia as its primary image suppliers for this title.

Marie Curie

Who Is Marie Curie?

Marie Curie loved to study the world around her. She is best known for her work with **radioactive** material. Today, this material helps fight **cancer**. This fight would not have been possible without Marie.

Marie was the first woman to win a Nobel Prize. This is an award given when someone changes the world with his or her work. Marie is the only woman to ever win it twice.

"Nothing in life is to be feared, it is only to be understood. Now is the time to understand more, so that we may fear less."

Marie is the only person to have won a Nobel Prize in two different sciences.

Early Days

Marie grew up in Warsaw, Poland. Both of her parents were teachers. Marie was the youngest of five children. She was taught to read and write at an early age. Marie had a very good memory. She worked hard at school.

When Marie was a child, Russia had control over Poland. Russia did not allow Polish people to read or write in Polish. Girls were not allowed to go to college. Marie's father lost his job. He also lost his savings. When Marie was old enough, she took teaching jobs to help her family earn money.

Marie had three sisters and one brother.

Where Is Poland?

Poland is a country in central Europe. It borders the Baltic Sea. Poland is known for its forests, rivers, and tall mountains. Warsaw is the country's capital.

Marie and Pierre Curie were married in 1895. They spent their honeymoon biking around France.

Starting Out

Marie wanted to continue learning. She also wanted to help her family. Marie worked to help pay for her sister to go to college in Paris, France. Six years later, Marie started school in Paris, too. Marie had read about **physics** and knew she wanted to be a scientist.

In Paris, Marie lived simply. She often ate bread and butter for meals. Marie studied hard. She also met her husband, Pierre Curie. They did **research** together. Both took teaching jobs to help fund their work.

When Marie was 15 years old, she graduated from high school. She won a gold medal for being a top student.

When Marie was **26**, she was at the **top of her class** in **physics** at her university.

Influences

While in Paris, Marie met many well-known scientists. One was Henri Becquerel. Henri had discovered radioactivity. This is the breaking up of part of an **atom**. It causes **rays** of energy to be released. Marie loved the idea of breaking down an atom. She became amazed with rays.

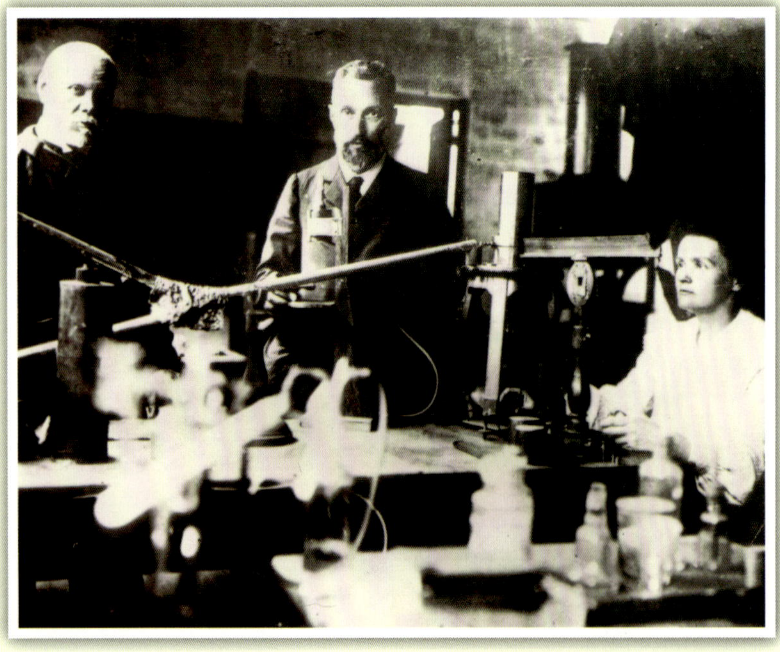

Marie, Pierre, and Henri often shared ideas and worked together.

Wilhelm Roentgen

In 1895, Wilhelm Roentgen discovered **x-rays**. Marie expanded on his work.

Henri Becquerel

Henri Becquerel worked with **uranium** salts. He noticed they were affected by light. The light gave off rays.

Pierre Curie

Pierre and his brother invented a new electrometer. This is a tool that measures very small electrical **currents**. Marie used this tool in her research.

Practice Makes Perfect

Marie began to experiment with rays. She tested a **mineral** called pitchblende. It was known to have uranium in it. She found that pitchblende gave off strong rays. The rays were stronger than those of uranium. Marie was very excited. She thought that pitchblende might have more **elements** in it.

Marie spent many hours in her **laboratory**. Each day, she and Pierre tried to find a new element in pitchblende. First, they discovered polonium. Later that year, they found another element. They named it radium.

During World War I, Marie made x-ray machines that could be moved. These "little Curies" helped more than **1 million** soldiers.

Marie and Pierre were great teammates. They spent their time working together on experiments.

What Is a Scientist?

A scientist is someone who studies things. Scientists who study energy are called physicists. Those who study substances are called chemists. Scientists are very curious. They love solving problems. Scientists try to answer questions through experiments. They collect information using a six-step method.

In 1910, after years of research, Marie was able to isolate pure radium. This means she was able to remove it from another substance.

Using a 6-Step Scientific Method

STEP 1

QUESTION

Scientists ask a question about what they want to learn. They read books and go online to research what other people know about the topic.

STEP 2

HYPOTHESIZE

The scientists then guess what the answer to their question might be. This guess is called a hypothesis.

STEP 3

EXPERIMENT

Scientists plan an experiment to see if their hypothesis is right. They gather the materials they need. Then, they set the materials up and do the experiment.

STEP 4

OBSERVE & RECORD

Scientists use their senses to observe what happens during their experiment. They watch. They smell. They listen, touch, and even taste. They then record what they find.

STEP 5

ANALYZE

Scientists think about what happened during their experiment. They decide if the experiment showed that the hypothesis was right or wrong.

STEP 6

SHARE RESULTS

Scientists let other people know about their experiment and what they found out. They may write a report or give a speech.

Pierre and Marie had two daughters, Irène and Ève. Irène went on to become a scientist as well.

Overcoming Obstacles

The Curies earned many awards for their work. Their discoveries also earned them money. However, working with radium was expensive. Even a small amount cost $120,000. Today, that would be millions of dollars. The Curies also had to live on the money they earned. They had to support themselves and their children.

Radium is a rare metal. It is silvery-white until it is exposed to air. Then, it darkens.

In 1906, Pierre died in an accident. Marie was suddenly a single mother. Although she was very sad, Marie continued working and researching.

Achievements and Successes

Countries around the world, including France and Poland, have created medals and coins to honor the work of Pierre and Marie.

Marie won Nobel Prizes in 1903 and 1911. The first prize was in physics. She won it for her work with rays. Her second prize was in chemistry. She won this prize for her work with radium.

Marie was given many other awards as well. She was honored in countries such as England, Italy, and the United States. These honors continue today. Both Poland and France declared 2011 the Year of Marie Curie. There are now two museums devoted to Marie and her work. One is in Warsaw, and the other is in Paris.

Today, people can visit the Marie Curie monument in Warsaw, Poland.

Marie ran a large laboratory at the Radium Institute. Scientists came from around the world to work with her.

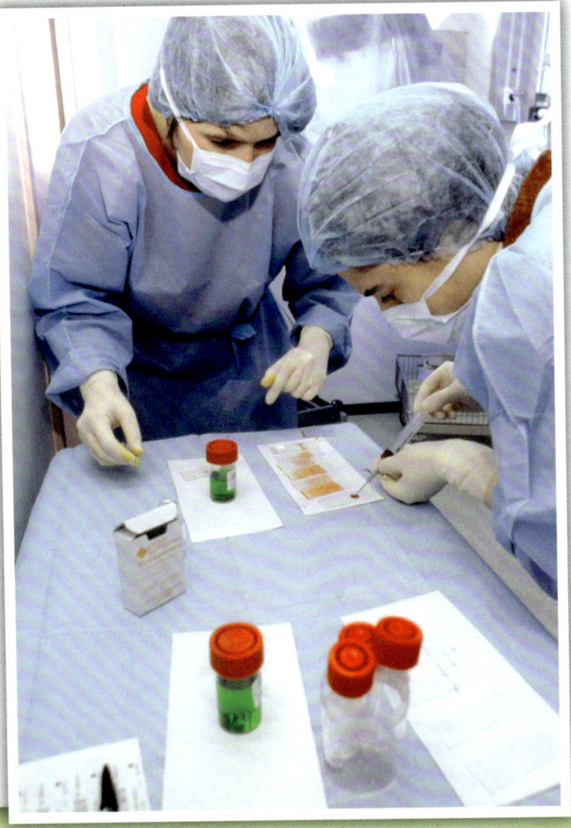

Researchers at the Curie Institute work with live cancer cells to find ways to fight the disease.

Impact on Society

At the Curie Institute, scientists work to understand and fight tumors.

Marie's research showed that cancer can be treated. She proved that cancer can be destroyed when it is exposed to radium. This process is called **radiation therapy**. It is key to the fight against cancer.

In 1914, the Radium Institute opened in Paris. The institute was built for Marie. It was to be a place for her to continue her studies. Six years later, the Curie Foundation was created. It helped raise money for the institute. Today, both are part of the Curie Institute. It is one of the world's leading cancer centers.

Timeline

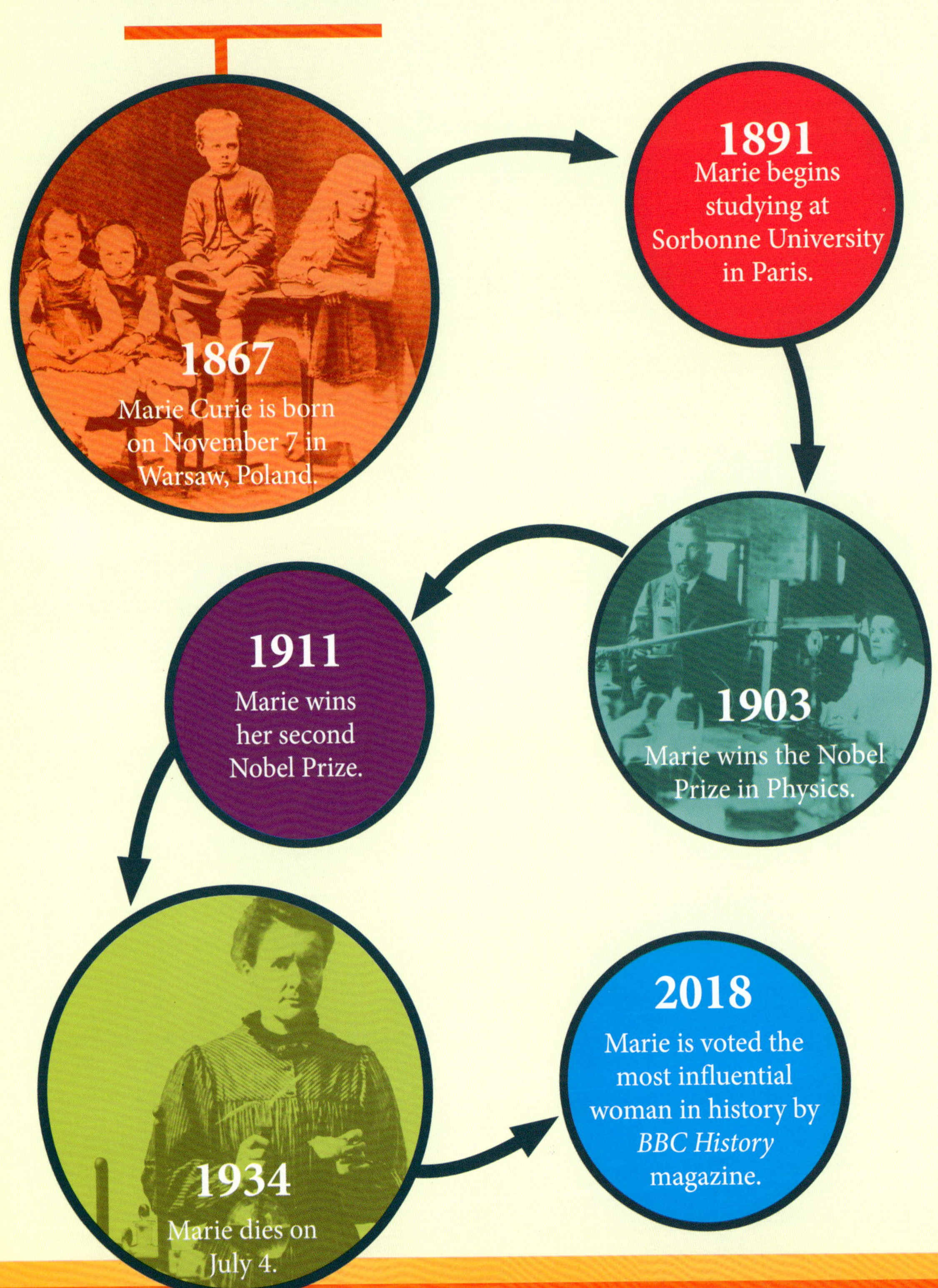
1867
Marie Curie is born on November 7 in Warsaw, Poland.
1891
Marie begins studying at Sorbonne University in Paris.
1903
Marie wins the Nobel Prize in Physics.
1911
Marie wins her second Nobel Prize.
1934
Marie dies on July 4.
2018
Marie is voted the most influential woman in history by *BBC History* magazine.

Key Words

atom: the smallest part of an element

cancer: a disease where abnormal cells grow in the body

currents: the flow of electric charges

elements: materials from which all other materials are made

laboratory: a room built for scientific experiments

mineral: a substance found in nature that is not a plant or an animal

physics: the branch of science that deals with motion, light, heat, sound, electricity, and force

radiation therapy: using beams of intense energy to kill cancer cells

radioactive: giving off radiation in the form of rays

rays: narrow beams of energy

research: studying materials to answer questions

uranium: a silvery substance found in certain rocks and soil

x-rays: a kind of radiation that can pass through substances that ordinary rays of light cannot pass through

Index

Log on to www.av2books.com

AV² by Weigl brings you media enhanced books that support active learning. Go to www.av2books.com, and enter the special code found on page 2 of this book. You will gain access to enriched and enhanced content that supplements and complements this book. Content includes video, audio, weblinks, quizzes, a slideshow, and activities.

AV² Online Navigation

Audio
Listen to sections of the book read aloud.

Book Pages
AV² pages directly correspond to pages in the book.

Video
Watch informative video clips.

Embedded Weblinks
Gain additional information for research.

Key Words
Study vocabulary, and complete a matching word activity.

Try This!
Complete activities and hands-on experiments.

Quizzes
Test your knowledge.

Slideshow
View images and captions, and prepare a presentation.

AV² was built to bridge the gap between print and digital. We encourage you to tell us what you like and what you want to see in the future.

Sign up to be an AV² Ambassador at www.av2books.com/ambassador.

Due to the dynamic nature of the internet, some of the URLs and activities provided as part of AV² by Weigl may have changed or ceased to exist. AV² by Weigl accepts no responsibility for any such changes. All media enhanced books are regularly monitored to update addresses and sites in a timely manner. Contact AV² by Weigl at 1-866-649-3445 or av2books@weigl.com with any questions, comments, or feedback.